edZOOcation

Cool Lizards

by Sara Karnoscak

ZOOKEEPER

Wildlife Tree
edZOOcation

Lizards don't smell with their noses. They smell with their tongues instead.

Do you smell with your tongue? What do you use to smell?

Most lizards can't walk on two feet. But frilled lizards can run on two feet!

Which animals can walk on four feet? Which animals can walk on two feet?

Some lizards are as short as an earbud. But some lizards are as long as a moving truck.

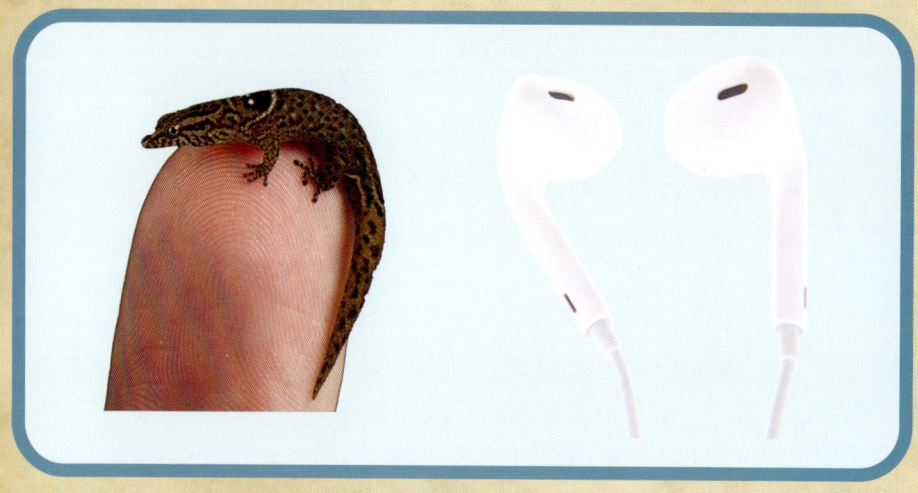

Which lizard is the shortest? Which lizard is the longest?

Most lizards can't change colors. But chameleons can change colors.

Do you see any chameleons? How many can you find?

Lizards don't have fur. They have scales instead.

Which animals have fur? Which animals have scales?

Most lizards have legs. But legless lizards have no legs.

Which lizards have legs? Which lizards have no legs?

Most lizards can
blink. But geckos
can't blink.

Which animals can blink? Which animals can't blink?

Some lizards can't climb walls. But some lizards can climb walls.

Which animals can climb walls? Which animals can climb trees?

Komodo dragons can't fly. But Komodo dragons can run fast.

Which dragons are real? Which dragons are pretend?

Most lizards sleep at night. But night lizards sleep during the day.

Which animals sleep at night? Which animals sleep during the day?

Goodnight, Lizards!

Dedication:

For London – fierce as a dragon, brave as a knight.

–S.K.

Eva and Moose, The Gecko

–A.R.

Author: Sara Karnoscak

Designer: Allyson Randa

Editor: Tess Riley

Photo Credits:

AdobeStock.com

Pixabay.com

Pexels.com

ISBN: 979-8-9894179-3-3

This book meets **Common Core** and **Next Generation Science Standards.**